1 MONTH OF
FREE
READING

at
www.ForgottenBooks.com

By purchasing this book you are eligible for one month membership to ForgottenBooks.com, giving you unlimited access to our entire collection of over 1,000,000 titles via our web site and mobile apps.

To claim your free month visit: www.forgottenbooks.com/free1056457

ISBN 978-0-331-25407-5
PIBN 11056457

U. S. DEPARTMENT OF COMMERCE
ROY D. CHAPIN, Secretary

BUREAU OF STANDARDS
LYMAN J. BRIGGS, Acting Director

CIRCULAR OF THE BUREAU OF STANDARDS, No. 400

INKS

[Issued December 30, 1932.   Supersedes C95]

UNITED STATES
GOVERNMENT PRINTING OFFICE
WASHINGTON : 1933

INKS[1]

ABSTRACT

The circular outlines briefly the history of iron gallotannate writing inks, gives formulas for three kinds of them, and has something to say about the aging of writing, and the restoration of faded writing. After this come brief discussions of several other kinds of inks, including colored writing inks, drawing, stamp-pad, recording, and other kinds. Formulas are given for most of them. Printing inks and others that depend upon pigments for their color and special properties are in a class by themselves. They are discussed, but no formulas are given for any of them.

The methods of testing in the Federal specifications are described.

In an appendix there are sections on weights and measures; on equipment for making ink in the home; and on dyes suitable for a variety of inks. Finally, there is a brief bibliography.

CONTENTS

[1] Prepared by C. E. Waters.

I. INTRODUCTION

Nobody can say how early in his long history man began to use signs and symbols to serve as reminders to himself, and to convey information to his fellows. No doubt the earliest of such signs were piles of stone, and the broken twigs we still use to mark an unfamiliar trail. The spirited though crude drawings left on the walls of European caves by the men of earlier cultures than ours show that primitive man was akin to us. Worse art is to be seen to-day on walls in public waiting rooms, especially along the lines of suburban trolley roads.

The walls of caves, flat rocks on the faces of cliffs, clay tablets, sheets of wax, and pieces of ivory, bone, and skin have all been used for writing upon. Even to-day a college diploma is a sheep-skin in name if not in fact, and tattooing has not died out. For centuries parchment was the material on which many books were written, but the papyrus roll was common enough to have given us the word "paper." These materials were expensive and could not be obtained in large quantities, and there could have been no great development of printing, or much letter writing, if paper had not been invented.

We may never know what was first used for writing ink. It may have been the juices of berries, or perhaps the colored earths that were stirred up with water and used as war paint.

About 1,200 years before the Christian Era the Chinese are said to have begun to make what is usually called India ink, by mixing soot with a solution of glue or of a plant gum. This pasty mixture was made into cakes which were dried. When some ink was needed for writing, the end of the cake was rubbed with water in a shallow dish. The nature of this ink made it necessary to write with a brush.

This circular has been written for the benefit of those who want to know something about inks and how to make them. The greater part of it is devoted to formulas and directions for making writing, hectograph, stamp-pad, recording, marking, and other inks. Printing ink and other inks that depend upon pigments for their color and their special properties are discussed, but no formulas are given for them.

The intent has been to give such detailed directions in most cases that any careful person will be able to make satisfactory ink. He may make a quart or less for his own use, but before he starts to produce ink on a commercial scale, as a newcomer in the field, he will be wise if he first considers the cost of establishing a profitable business, selling a cheap article with an unfamiliar label.

II. IRON GALLOTANNATE INKS

1. ANCIENT INKS

Leather tanned with bark was known before the Christian Era, and the staining of wet leather by contact with iron must have been noticed often. Yet the world waited for more than 2,000 years after the discovery of India ink, or until about 1126 A. D., before tannin and iron were combined to make writing ink. This kind of ink is still used in larger quantities than any other. The ink was made by dissolving ferrous sulphate (copperas or green vitriol) and glue in an infusion of nutgalls, which contains a kind of tannin that is especially suitable for making ink.

The infusion of nutgalls was allowed to ferment, the other materials were added, and the mixture allowed to stand for a time. When it was quite black, it was used as ink. The change in color was caused by the action of oxygen from the air upon the iron salt. Ferrous iron forms with tannin an easily soluble compound that is not intensely colored, but the ferric compound formed by the oxidation is black and nearly insoluble in water. Ink made by this process was a muddy fluid in which floated innumerable microscopic particles of the black ferric compounds. The glue, or the plant gum if it was used instead, helped to keep the particles from settling to the bottom of the fluid, and later served to fasten them to the paper or parchment.

In those days there was no thought of chemical control of the manufacturing process, nor any chemist who could have supervised it. Not until 1748, when William Lewis began to experiment, was any attempt made to produce a balanced ink, with nearly correct proportions of iron and nutgalls; and even in his time there were no analytical methods to help him. Though he had to work by the cut-and-try method, he tried.

As each ink maker used the formula he considered the best, but had no idea of the amount of tannin in the galls, or of the purity of his ferrous sulphate, many a batch of ink must have been far from balanced in composition. This state of affairs is reflected by the condition of various old documents preserved in European libraries and elsewhere. The writing of some is still legible, and the parchment or paper in good condition. In others the paper is more or less eaten through by the ink, which, it is generally believed, contained too much sulphuric acid, which does not evaporate, and which has a sort of charring effect upon paper. Another explanation is that iron oxide formed from the ink destroyed the paper in the same way as a rusty nail attacks a piece of wood. Whatever the real cause, it is said that in extreme cases only the unwritten margins of the pages are left, for the rest of the paper has crumbled away.

2. MODERN INKS

Early in the nineteenth century a change was made in the manufacture of writing ink. Instead of deliberately allowing it to oxidize and be turned into a muddy fluid, it was guarded from the action of the air and kept clear as long as possible. When a batch is made, nowadays, it is allowed to remain undisturbed for a time so that solid impurities will settle to the bottom, but only a small part of the iron salts in a vat containing some hundreds of gallons of ink will become oxidized.

The coloring matter in the older inks consisted of black particles that remained to a great extent upon the surface of the paper. The clear inks soak into the fibers of the paper, or between them, and then become oxidized. For this reason it can be argued that the clear inks should be the more permanent, because so little of the writing is on the surface where it can be rubbed off. To keep the ink clear as long as possible, it must be kept from oxidation, and must also contain a small quantity of free hydrochloric or sulphuric acid. The more free acid the ink contains, the longer will it remain clear, but the greater will be its destructive effect upon paper, and its corrosive action on steel pens. There must be some sort of compromise if the use of iron gallotannate (or iron-gall) ink is not to be abandoned. Our ancestors a few generations back were not concerned with the acidity of their ink. It was muddy anyhow, and they had no steel pens to be corroded. The fountain pen with its noncorrodible point puts a temptation in the path of the ink maker, who knows what an extra amount of acid will do for him in keeping the ink clear.

Ink which has undergone but little oxidation does not look intensely black in the bottle, and makes such pale marks on paper that it is necessary to give it a stronger color by the addition of a dye. The dye would not be needed if the writer could keep his letters for a day or two for them to become easy to read. When clear gallotannate inks began to be made, synthetic or so-called aniline dyes were something yet to be discovered. Of the comparatively few available dyes in those days, it is probable that only indigo could be used without causing the precipitation of solid matter in the ink. Indigo itself is not soluble, but by suitable treatment with strong sulphuric acid it is converted into the disulphonic acid, which dissolves readily and forms no precipitate by combining with the other ingredients of the ink.

3. A DEFINITION OF INK

In 1890, Schluttig and Neumann, ink chemists of Dresden, Germany, wrote what is in many respects the most important book on iron gallotannate inks, because of its far-reaching and lasting influence.[2] Their definition of ink, their explicit recommendations for making record ink, and the whole tone and spirit of the book set a new mark for the ink manufacturer to aim at. As a basis for some of the discussion in the pages which follow, their definition, in nearly literal translation, is here given.

By ink we mean a liquid, suitable for writing, which—
1. Is a clear, filterable solution, not a suspension.
2. Is mobile and keeps for a considerable time; that is, it flows easily from the pen, and neither clogs, drops off, nor spreads on the paper.
3. Has good keeping quality in glass; that is, in the inkstand it forms—
(a) A slight deposit slowly.
(b) No skin-like deposit, on the surface or on the walls, and never any mold.
4. On a good pen it forms only a slight, varnish-like, smooth coating, but not a loose, crusted one.

---

[2] Schluttig, O., and Neumann, G. S., Die Eisengallustinten (iron-gall inks), v. Zahn & Jaensch, Dresden, 1890.

5. Has no pronounced odor.

6. Is not too acid and does not penetrate through good paper.

7. Has an intense color, which does not become paler nor bleach out entirely in the liquid or on paper (in the latter case judged after the complete drying of the writing, for moist lines always look darker than dry ones).

8. Gives writing that is not sticky after drying.

9. Gives writing that, after drying for eight days, is not removed by water or alcohol—even by treatment for days—to such an extent that it becomes illegible.

10. Has a definite minimum content of iron.

11. And contains enough tannin; that is, it gives writing which after drying becomes deep black within eight days, and which, even after treatment for days with water and alcohol, still retains a certain degree of blackness.

All the points of the definition are important, but not equally so. The authors stressed 1, 3, 6, 10, and 11. In the work described in the book they attempted to make ink that met the requirements of their definition to the fullest possible degree. Taking it for granted that the ink should contain iron, they first studied the effect of using different amounts of that metal and of gallic and tannic acids in the solution. Having done their best with these materials, they extended their investigation to include inks made with iron and substances that are closely related to gallic acid in their chemical structure. Their conclusions may be better understood after a brief discussion of gallic and tannic acids.

4. THE TANNIN IN IRON GALLOTANNATE INKS

The tannins are a group of more or less closely related substances that are found in many different kinds of plants. Their name comes from their use for tanning the hides of animals to make leather. The chemistry of this group of substances is quite complicated, and only a start has been made in determining the molecular structures of the different tannins. It has been shown that at least some of them are glucosides, or compounds of the familiar sugar, glucose (dextrose), with varying amounts of gallic acid, digallic or tannic acid, and possibly some trigallic acid. Chemically, glucose is an alcohol, and its compounds with these acids are esters, or salts, as truly as ethyl acetate is the ester or salt of ethyl alcohol and acetic acid. The sugar and the acids can be separated by the same type of reaction that splits apart ethyl alcohol and acetic acid from ethyl acetate.

Gallic acid is 3, 4, 5-trihydroxybenzoic acid, and therefore has the structural formula I. Tannic, or digallic, acid is galloyl gallate,

and formula II is assigned to it. It will be seen that each of these acids has a free carboxyl (CO.OH) group, characteristic of the majority of organic acids, and this makes it possible for the acids to combine with glucose.

In addition to these acids, Schluttig and Neumann used 26 other substances that differed from them in the number and arrangement of the carboxyl and hydroxyl (OH) groups in the molecule, or that had methoxyl (OCH_3) groups instead of hydroxyl. Their final conclusion was that in order to be suitable for making ink of good color and permanence, the tannin must contain three adjacent, free hydroxyl groups. This condition is satisfied by tannic and gallic acids, and these of all the substances studied were found to be the best for making ink. Thus a formula probably discovered by accident, and improved empirically during the centuries, is shown to be scientifically correct.

5. FORMULAS FOR IRON GALLOTANNATE INKS

(a) HISTORY OF GOVERNMENT RECORD INK FORMULA

It does not suffice to find out the best materials for making ink, for unless they are used in the correct amounts the ink will not be good. There should be no excess of either iron or tannin, the amount of free mineral acid should be just enough to keep the ink clear for a reasonably long time in the bottle, and there must not be a deficiency or excess of dye. The formula for " copying and record " ink, given on page 7 of this circular, differs in but two respects from the one recommended by Schluttig and Neumann as the result of their investigation. They used gum arabic, at the rate of 10 g per liter, to act as a preservative. As we would now say, it served as a protective colloid to hinder the precipitation of any insoluble ferric gallotannate that might be formed in the ink. Some years ago the gum was omitted from the United States Government formula by a clerical error, which was discovered too late to be corrected, because the contract for a year's supply of ink had already been awarded. The Bureau of Chemistry, Department of Agriculture, then made some special tests which showed that the omission of the gum was an improvement, and ever since the gum has been left out of the formula for the standard ink.

The Government formula calls for a dye that is different from the one recommended by Schluttig and Neumann, though the two are closely related, chemically. This is further discussed on page 36.

The Commonwealth of Massachusetts adopted the Schluttig and Neumann formula as the official ink for records and other public documents, in the early 1890's, a few years before the Federal Government started to use it. Connecticut followed their lead about 1914. It should be noted that Prussia, in 1912, decreed that the official ink should contain at least 4, and not more than 6, g of iron per liter, these being the limits set by Schluttig and Neumann for record ink. In this country the maximum, 6 g of iron, has always been required. This weight of the metal is contained in the 30 g of ferrous sulphate crystals in the formula. Massachusetts requires the use of gum arabic, but curiously enough is silent about the use of blue dye.

(b) STANDARD FOR GOVERNMENT COPYING AND RECORD INK

In 1924, the Federal Specifications Board took over the old specification for " Treasury standard " writing ink, and promulgated it as United States Government Master Specification No. 163, Record and Copying Ink. It was published as Bureau of Standards Circular

No. 182 for some years, but it is now out of print. Although not a true copying ink, it will give one good press copy when the writing is fresh, and this generally suffices. In 1930, the specification was changed in form, but not in technical requirements, and issued as Federal specification TT–I–521, Ink, Copying and Record, a part of the Federal Standard Stock Catalogue. The apparent subordination of "record" to "copying" is the result of alphabetical exigencies, so that the specification will fit into its proper place in the catalogue.

Like other specifications for inks, this one gives a formula for making ink to be used as a standard for comparison in testing samples of inks bought by the departments. Many who read the specification labor under the mistaken impression that the manufacturer must use the same pure materials for producing the ink he furnishes. This idea is not in accord with the wording or intent of the specification. It is necessary, in order that anybody can make the same standard ink, to have all the ingredients of definite purity. The manufacturer who possesses the knowledge and skill to use cheaper raw materials in making ink that meets the requirements of the specification may do so.

A standard ink is necessary because some of the requirements can not be stated in exact terms, and some of the properties of the ink as measured by the tests may differ somewhat according to the conditions. If the standard and the sample are put through the same series of tests, side by side and at the same time, it is easy to see whether the sample is equal to the standard in all essentials.

In the following formula for the standard copying and record ink of the specification, all of the materials must be "of the strength and quality prescribed in the edition of the United States Pharmacopœia which is current at the time bids are asked for."

	Grams
Tannic acid	23. 4
Gallic acid crystals	7. 7
Ferrous sulphate	30. 0
Hydrochloric acid, dilute, U. S. P	25. 0
Carbolic acid (phenol)	1. 0
Soluble blue (C. I. 707; Sch. 539)[3]	3. 5
Water to make a volume of 1 liter at 20° C. (68° F.).	

(c) FORMER STANDARD FOR COPYING INK

There is no Federal specification for true copying ink. Years ago fairly large quantities of "Treasury standard" copying ink were bought on a specification based on the following formula:

	Grams
Tannic acid	46. 8
Gallic acid crystals	15. 4
Ferrous sulphate	60. 0
Hydrochloric acid, dilute, U. S. P	50. 0
Gum arabic	10. 0
Carbolic acid (phenol)	1. 0
Suitable blue dye[4]	5. 0
Water to make a volume of 1 liter at 20° C. (68° F.).	

---

[3] The symbols in parenthesis, here and throughout this circular, indicate definitely the dye type intended. See also p. 34.
[4] In those days the dye was generally expected to be Bavarian blue DSF, as recommended by Schluttig and Neumann. See also p. 36.

To improve their copying qualities, some inks are made with dextrin, sugar, glycerol (glycerin), or other similar substance. If too much is used the writing will be sticky.

(d) STANDARD FOR GOVERNMENT WRITING INK

The copying and record ink is of too heavy a body to please most writers, so there is a Federal specification, TT–I–563, Ink, Writing. It was written originally to provide ink for use in post-office lobbies, where the conditions are devastating to pens. The standard ink of this specification is similar to some of the commercial writing fluids. Except for the amounts of dye and preservative, it is half as concentrated as the standard copying and record ink. The effectiveness of the preservative (phenol) depends upon the quantity of it in a given volume, and the solution must contain enough dye to give a good color to the fresh writing; so the weights per liter of these two ingredients are the same as in the more concentrated ink. The formula for the standard writing ink is:

	Grams
Tannic acid	11. 7
Gallic acid crystals	3. 8
Ferrous sulphate	15. 0
Hydrochloric acid, dilute, U. S. P	12. 5
Carbolic acid (phenol)	1. 0
Soluble blue (C. I. 707; Sch. 539)	3. 5
Water to make a volume of 1 liter at 20° C. (68° F.).	

It is hardly necessary to say that the materials used for making this ink must be of the same quality as those in making the other standard iron gallotannate inks.

(1) *Concentrated ink.*—Concentrated ink is accepted if it meets the requirements of the specification for writing ink. The contents of the usual small bottle or collapsible tube in which the ink is packed will make a quart of writing fluid when mixed with water. Hydrochloric acid is a solution of a gas in water, and it is therefore completely volatile. For this reason it is probable that most producers of concentrated ink use an equivalent amount of sulphuric acid instead. This will be 1.77 g, or less than 1 ml, of sulphuric acid of specific gravity, 1.84.

Concentrated ink occupies less space and weighs less than its equivalent in writing fluid, so the bottle is less apt to be broken in shipment, to say nothing of the saving in express or postal charges. It is also less apt to freeze and burst the bottles than the more dilute writing fluid; while if breakage should occur, surrounding packages will suffer less harm.

(2) *Ink powders and tablets.*—Ink powders and tablets represent the last step in concentrating ink. At the time the specification for writing ink was prepared, and for some years afterwards, the few samples of these materials that had been examined by the Bureau of Standards consisted wholly of dyes, or else they were quite unsatisfactory mixtures that purported to make good iron gallotannate ink. It is needless to point out their faults in detail. The chief problem to be solved in making an ink powder—a tablet is only the compressed powder—is to find a dry acid that will fully take the place of hydrochloric or sulphuric acid. This problem was solved in 1931 by one manufacturer whose product was given a severe test. The powder was not analyzed by the bureau, however.

6. PREPARATION OF STANDARD IRON GALLOTANNATE INK

In the formulas for the iron gallotannate inks, and elsewhere throughout this circular, when water is called for, it is to be understood that distilled water is best, with rainwater the second choice. There are parts of the country where the stream and well water is so hard, because of the calcium carbonate dissolved in it, that a substantial portion of the acid in iron gallotannate ink will be neutralized if the water is used for making the ink. Water containing 500 parts per 1,000,000 of calcium carbonate, or 0.5 g per liter, is not unknown. This quantity of " lime " will neutralize about 3.5 g, or nearly one-seventh, of the dilute (10 per cent) hydrochloric acid in the copying and record ink. As the amount of acid is cut to a minimum in the formula, it is evident that the quality of the ink will suffer if so much of the acid is neutralized. It will deposit sediment prematurely. Writing ink contains only half as much acid as the other, so the hard water will neutralize almost 28 per cent of the free mineral acid. Water that has been softened by the zeolite, or base exchange, method is as alkaline as before the treatment, and will neutralize as much of the acid.

To make a liter of any of the standard inks, dissolve the tannic and gallic acids in about 400 ml of water at about 50° C. (122° F.). In a separate vessel dissolve the ferrous sulphate in 200 ml of water which contains the requisite amount of hydrochloric acid. In a third vessel dissolve the dye in about 200 ml of water. Mix the three solutions in a measuring flask, rinse the vessels with small portions of water, and use the rinsings to make the total volume up to 1 liter, after cooling to 20° C. (68° F.). The carbolic acid can be added before the final adjustment of the volume, if it is thought necessary to be so exact. The ink should be mixed very thoroughly by inverting the stoppered flask at least a dozen times, before any is removed.

Hydrochloric acid is a gas which dissolves readily in water. The usual concentrated acid contains about 36 per cent by weight of the gas, the rest being water. The Pharmacopœia defines " dilute " hydrochloric acid as containing not less than 9.5, nor more than 10.5 per cent of the gas. The plain intent is to make 10 per cent acid, and directions for preparing it are given. There are reasons for suspecting that some who make ink do not understand how to prepare the dilute acid, and that they regard the concentrated acid as 100 per cent. If, acting on this belief, they mix 10 parts by weight of the acid with 90 parts of water, they will have 3.6 per cent acid. The correct way is to mix 100 parts by weight of the concentrated acid with 260 parts by weight of water. The 360 parts of the mixture will contain 36 parts of hydrochloric acid gas, or 10 per cent.

The same principle must be applied if the strong acid is of some other concentration than 36 per cent. If it seems preferable to measure the acid and water instead of weighing them, the density of the acid must be taken into account. The density of the 36 per cent acid is about 1.19. If 100 ml is taken, it will weigh 119 g, and will contain 42.8 g of hydrochloric acid gas, the last figure being 36 per cent of 119. To get 10 per cent acid, add 309 ml of water, which will weigh the same number of grams, giving a total of 428 g of the dilute mixture.

Some manufacturers claim that sulphuric acid is better than hydrochloric. One part by weight of hydrochloric acid gas is chemically equivalent to 1.345 parts of 100 per cent sulphuric acid, or to 1.416 parts by weight of the usual concentrated acid of 95 per cent strength (66° Baumé; density, 1.84). Hence the equivalent of 25 g of dilute hydrochloric acid containing 2.5 g of the gas is $2.5 \times 1.416 = 3.54$ g of ordinary concentrated sulphuric acid.

An ideal but quite unpractical way to keep iron gallotannate ink is in glass globes hermetically sealed by melting the glass together. A sample preserved in this way had only a little sediment in it after being kept for five and one-half years. As the ink must be kept in bottles, these should be nearly full, and the corks should be as perfect as possible, and not the inferior ones with numerous cavities through which air can get in to the ink. It is best to keep the bottles in the dark.

In the appendix is a brief discussion of dyes that can be used in iron gallotannate ink.

7. AGING OF WRITING

The behavior of iron gallotannate inks on paper is so important that it deserves to be discussed in some detail.

The fresh writing is blue, except in the rare case of the ink containing a black, instead of the usual blue, dye. In a few hours the writing becomes perceptibly darker because the ferrous salt in the ink has begun to oxidize to black ferric gallotannate. Under ordinary conditions of diffused daylight the writing should attain its greatest intensity of color, a deep blue-black, in about a week. If the ink is unusually acid, the color develops more slowly. On the other hand, if the ink contains too little acid, or if the acid is neutralized by exposing the fresh writing to the fumes of ammonia, the blackening will take place in a day or two.

The oxidation that causes the blackening of the writing does not cease abruptly when all of the ferrous iron is converted into ferric iron. The dye and the gallic and tannic acids are also subject to oxidation. In the course of time the dye will disappear. If this occurs before the two acids have been affected to any great extent, the writing will still be black, but no longer blue-black. If the ink maker depended more upon dye than upon iron gallotannate, it does not seem impossible that the aging writing will never go through the true black stage, but when all the dye is gone, a substantial part of the gallic and tannic acids will have gone with it, and then the writing will have a brownish color. It is certain that if the paper endures long enough, finally nothing will be left of the writing but rusty lines of ferric oxide on the paper.

It is natural to ask how long it takes for all this to happen to the writing, and the reader can draw his own conclusions after being reminded of some of the factors which influence the rate at which the ink ages. First of all comes the ink itself. If it is truly a record ink, the writing ought to last for centuries if it is kept under proper conditions. Ordinary writing ink can not be expected to last as long as record ink, and if both are used in the same document, as when two persons sign it, each with his own fountain pen, one might draw the conclusion 20 or 30 years later that one signature is much more recent than the other. The fluidity of the ink and the absorptiveness of the paper influence the quantity of ink in the written characters, and thus play their part in the aging. If the writing was blotted instead of being allowed to dry naturally, the ink is handicapped at the start. Inspection of old notebooks, ledgers, and similar records is apt to disclose considerable differences in the appearance of writing of the same date and in the same hand. Such differences are outstanding in the notebooks to be discussed further on.

The appearance of the writing after a number of years have gone by, will also depend upon how it has been kept. If it has been much exposed to light and has been in a damp place, the writing will look older than if it has been kept in a dry atmosphere and in the dark. Apparently the character of the paper is not to be disregarded. According to Schluttig and Neumann, fresh writing will darken much more rapidly on some kinds of paper than on others. It is not to be supposed that this hastening of the darkening by the paper, that is, its oxidizing effect upon the ink, will cease in a short time. It seems more reasonable to think that the paper will continue to affect the writing over a period of years. If it does, then of two pieces of writing, identical in every respect except the kind of paper, one will age more quickly than the other.

As regards the permanence of the blue dye in writing with iron gallotannate ink, an account of a simple test made on dated entries in some laboratory notebooks may be of interest. The books have been kept together in a drawer under good conditions for preserving the writing. The results do not support the common belief that within 15 years all the blue dye will have disappeared.

During the years to be considered, 1905 to 1910, many new supplies of ink must have been used in the fountain pen. Probably most of it was " Treasury standard " ink, and for that reason it is likely that the pen was once in a while replenished with water, instead of with the heavy-bodied ink. It is also probable that more than one manufacturer supplied it. Some of the writing was blotted, and some allowed to dry spontaneously, as indicated by the varying degrees of blackness of writing on the same page. In brief, these entries are in reality a miscellaneous lot of writing, though all in the same hand.

Samples of the writing that looked about equally black, and scattered among various dates from April, 1905, to July, 1910, were selected. To each a small drop of distilled water was applied, allowed to soak in for about 10 seconds, and was then pressed for the same length of time with a piece of white filter paper, held down firmly with the tip of a finger. The filter paper was then examined for signs of blue dye " bleeding " from the writing. In all, 22 samples of writing were tested in this way. In the case of 2 samples in 1905 and 1 in 1910, one could not be sure whether there was any bleeding of dye. In 10 other cases, 2 in 1905, 1 each in 1906 and 1907, 3 in 1909, and 3 in 1910, there was no evidence of the presence of blue dye in the writing. In the remaining 9 cases, 1 in 1905, 3 in 1907, 2 in 1908, and 3 in 1910 showed more or less staining by the dye. The faintest stains were 1 each in 1905 and 1908, and

2 in 1910. It is evident that these results do not support the claim that all the dye disappears within 15 years. Like other matters of opinion relating to the age of writing, this lends itself admirably to forensic disputation.

Comparatively fresh writing is readily acted upon by such chemicals as oxalic acid, caustic soda, and ink eradicator. If so mild and simple a chemical reagent as distilled water will bring out such marked differences as those described in the preceding paragraph, one should be cautious in drawing inferences from the action of other chemicals upon writing of unknown age.

8. RESTORATION OF FADED WRITING

The choice of iron gallotannate ink for records is generally based upon its permanence, without much regard to another advantage it has over an ink made by dissolving a dye. When a dye fades, it sometimes leaves literally nothing behind on the paper, though it may leave traces of oxidation products that can not be detected readily, if at all. When an iron ink fades away, it leaves small quantities of iron oxide in the pen strokes, and this makes it possible to restore the writing.

If the paper is first steamed to moisten it slightly, the vapors of ammonium sulphide will change the iron oxide into iron sulphide, which is brown or black, according to the quantity. The color change is not permanent, and there may be bad aftereffects upon the paper. A 2 or 3 per cent solution of tannic acid will bring out the writing in the same way. Again, a slightly acidified solution of potassium ferrocyanide will convert the iron oxide into Prussian blue. The treatment with these two solutions is best accomplished in a letter press, by placing cloths or pieces of white blotting paper moistened with the solution in contact with the faded writing, and keeping the whole under pressure for a few minutes.

A disadvantage of any chemical treatment is that iron is to be found in practically everything. Paper and dust contain it, and if the document has been much handled, it will have a surface coating, containing iron, that comes from dirty hands. The chemicals that are applied for the purpose of restoring the writing are not selective, but also act upon the iron from other sources, so that when the job of restoration is done, the writing will appear as dark lines upon a less dark, and probably unevenly colored, background.

If a source of ultra-violet radiation is at hand, it is better to defer the application of chemicals until the effect of this radiation has been tried. Under the right conditions, the iron oxide will be made to glow with a phosphorescent light that can be photographed. It is thus possible to get an exact copy of the writing without running the risk of ruining the document permanently.

A number of years ago, when it was decided to place the Declaration of Independence on exhibition in the Library of Congress, the Bureau of Standards was consulted about the advisability of intensifying the faded writing by chemical treatment. After serious consideration of the uncertainty of success, and the danger of the action of the chemicals upon the document, it was decided that it would be better not to tamper with it, at the risk of ruining it forever.

III. VARIOUS KINDS OF INKS

1. CARBON INKS

(a) CARBON WRITING AND DRAWING INKS

Carbon can not be bleached by any amount of exposure to intense light, and it resists attack by chemicals that quickly destroy paper. If carbon could be dissolved in water, it would be ideal material for making black writing ink. India ink, which has already been mentioned briefly, is not a solution, but a suspension of carbon in water containing glue or gum. The Chinese, who writes with a brush, does not care if the carbon settles to the bottom of his little saucer of ink, for he can stir it up at each dip of his brush. To us his ink and his brush would be intolerable, and fortunately for those who must use carbon ink, we have learned how to keep the carbon from settling to the bottom.

If some lampblack is stirred with water and then left to itself, after a time it will settle and leave clear water above. If the mixture is ground for a long time and then allowed to rest, the carbon will not settle so quickly as before, and part of it may be very slow, indeed, in reaching the bottom. If instead of pure water, a solution of some plant gum, of shellac and borax, or of soap is used, after thorough grinding the carbon will tend to remain in suspension. The difficulty in preparing a permanent suspension of carbon lies in the grinding. If it is ideally complete, no two particles of carbon would touch one another, much less cling together, but each would be separate from all others, and each would be coated with an adsorbed film of gum, shellac, or soap. It would then have almost no tendency to settle.

The manufacturers of black drawing ink are very successful in making practically permanent suspensions of carbon. According to one manufacturer, the ink mixture is ground for three or four weeks in a ball mill. If this is necessary, it is a sufficient reason why carbon ink can not be made satisfactorily by hand grinding.

Water that is clouded by extremely fine particles of clay in suspension will clear quickly if some salt is dissolved in it. The carbon in ink will soon settle if some acid is added, while alkali makes the suspension more stable. The ammonia that can be smelt in some kinds of drawing ink is a mild alkali. Because of the sensitiveness of carbon suspensions to acids, carbon inks can not be mixed with iron gallotannate ink, and a fountain pen that has held the latter must be cleaned with extreme care before filling it with carbon ink. Carbon writing inks might be popular if it were not so easy to ruin them by ignorance of their peculiarities, or by carelessness.

(b) PRINTING, CANCELING, AND OTHER CARBON INKS

Black drawing ink contains only a small percentage of solid matter, and it does not differ greatly from clear inks in fluidity and working qualities. The situation is different with some of the other kinds of carbon inks. The carbon in a canceling ink should be carried into the paper, and remain there in spite of attempts to remove the marks by washing. Mimeograph and other duplicating machine inks also require great care in their formulation and manufacture. If the mimeograph ink is not just right, the copies may be too pale, or blurred from too much ink; the carbon may clog the stencil, or become smeared over its face.

Printing inks contain more carbon than any other kind of ink, and among themselves they differ widely. Some are thick liquids, and others are stiff pastes, with all consistencies between. It is necessary to adapt the physical properties of the ink to the kind of printing to be done. The same ink can not give equally good results in printing from ordinary type, from a lithographic stone, a halftone cut, and an engraved plate; and the paper introduces another important factor in the results.

Formulas for various kinds of carbon inks are to be found in books, but they should be regarded as only suggestions. The character of the finished ink depends upon the physical and chemical properties of the ingredients, as well as upon the amount of each used; and the way the ink is made is of importance. There are no Federal specifications for any of these inks, because there are no laboratory tests that can take the place of actual trials on the press, and with the paper and the kind of work for which the ink has been made. It requires skill and experience to make good pigment inks. The manufacturer has his working formulas, but he would not turn over the actual production to an unskilled person.

It is possible to measure some of the properties of some of the ingredients of printing ink, but there are no adequate tests which enable one to predict exactly what the finished ink will be like. The consistency, for instance, depends to a great degree upon what is called the oil absorption of the pigments. This differs according to the chemical composition, and is closely tied up with the fineness of grain.

The difficulties involved in the manufacture of a pigment ink are well illustrated in the description of the complicated procedure followed by the Government Printing Office in order to make acceptable mimeograph ink.[5]

2. DYE INKS FOR WRITING

As explained on an earlier page, when an iron ink fades from age, it leaves behind at least a little iron oxide, and thus it is possible to reveal the writing by suitable chemical or other means. When the ink is a solution of a dye, there is no such possibility of restoration when the writing fades. Oxidation of the dye forms volatile products which escape into the air, or maybe small amounts of other substances which remain in the paper, but with which there is no certain way of forming colored compounds. For these reasons dye solutions are not regarded as suitable for record inks. On the other hand, they have advantages over iron gallotannate inks. They keep almost indefinitely in the bottle, are seldom corrosive, and because they are much more dilute than iron inks, they do not form thick deposits if they dry on pen points. A liter of standard writing ink contains a total of 35 g of iron and other nonvolatile solids, while the same volume of a dye ink will contain perhaps as much as 10 g.

---

[5] Government Printing Office Tech. Bull. No. 15, Standard Mimeograph Ink and Paper.

In Federal specification TT–I–549, Ink, Red, the standard is made by dissolving 5.5 g of crocein scarlet MOO (C. I. 252; Sch. 227) in 1 liter of water. A still weaker solution of methyl violet B (C. I. 680; Sch. 515) would suffice for that color.

There are no Federal specifications for dye inks of other colors than red. There are many water-soluble dyes, and it is possible to make inks of almost any shade and hue by dissolving suitable dyes in water. If it should turn out that a particular ink has a tendency to "feather," or make blurred spreading lines on paper, this can be prevented by dissolving in the ink some gum arabic at the rate of 20 or 30 g per liter.

Many dyes have an antiseptic action, so their solutions do not become moldy, though no preservative is added to them. With other dyes it is necessary to add about 1 g per liter of phenol or other suitable preservative. In the case of a dye that is just on the border line, the presence of gum arabic might encourage the growth of mold that would not otherwise thrive.

Now and then somebody wants to be told how to make washable ink. In the generally accepted sense, a washable ink would be one that is not removed by washing; in other words, it is indelible ink. The inquirers, however, use the word in exactly the opposite sense, for they want to make ink that can be washed out of fabrics easily and completely. No study of this subject has been made by the bureau, but in theory it is easy to make such inks, by selecting dyes that do not fix themselves upon the fabric except with the aid of a mordant, or substance that forms an insoluble compound with the dye. Assuming the fabric to be made of cotton or other vegetable fiber, one need only avoid the direct, or substantive, dyes, for these require no mordant. If the fabric is silk or wool, the problem is far more difficult, because these fibers can be colored with almost any class of dye.

Although dye inks are not considered suitable for records, they are not to be condemned on that account. They are excellent for ordinary correspondence, and for writings that are to be kept for only a few years. If preserved in a dry place, and not in the light, there is no reason why writing with dye ink should not last for several decades. At the Bureau of Standards is a book in which there are several press copies of letters written early in 1901. The inks represented are iron gallotannate, blue from a "copy blue" typewriter ribbon, violet from printing ink containing dye of that color, and another violet from rubber-stamp ink. There are also two red lines on a page with a drawing. Evidently the drawing was made on a scratch pad, and when the sheet was torn off in order to make the press copy, a little of the red glue ("padding compound") that was on two edges of the pad, came with it. This was enough to make the two red lines in the press copy.

A press copy contains only part of the coloring matter of the original writing, and it might be expected to fade quickly on that account. In spite of this, the copies are still of good color, though 31 years old when this was written, and there is nothing to indicate that they will not last as long again, if kept under the same favorable conditions as heretofore.

3. PRUSSIAN BLUE INKS

Prussian blue is not a dye, but it has as great coloring power as some dyes. It is ordinarily quite insoluble in water, but a variety known as soluble Prussian blue can be made. It does not form a true solution, as salt and sugar and many other substances do. It is more like the suspensions of carbon and clay that have been mentioned, but its particles are so small that the suspension, or colloidal solution, looks perfectly clear. The colloidal solution is easily disturbed by salts. The blue is first formed as an insoluble precipitate, which must be washed repeatedly until all the salts it contains are removed. This is most easily done by dialysis. The washed precipitate will run through any filter, so the bag of cellophane or parchment paper is hung in a warm place for the contents to dry. If properly made, the blue can be dried and kept indefinitely, and a piece placed in distilled water will quickly make a clear blue solution.

It has been found that a mixture of Prussian blue and oxalic acid will give a clear, stable solution. In various published formulas the quantity of oxalic acid called for varies from one-fourth to one-half the weight of Prussian blue. In making a few experimental solutions at this bureau, 1 part of acid to 5 of blue was found to be enough. Writing done with Prussian blue ink is very fast to light and water. It can be removed, though not easily, by treatment with the usual ink eradicator that comes in two bottles, one containing an acid, generally oxalic, and the other a hypochlorite solution. It is sold commonly as "acid-proof ink," but alkaline solutions, even soap and water, decompose the Prussian blue and leave a rusty stain of hydrated iron oxide, which can be removed by further washing, or by treatment with dilute acid.

A solution of 10 g of Prussian blue and 2 g of crystallized oxalic acid in a liter of water makes writing fluid of a good depth of color, though some may prefer a stronger solution. As bright blue ink is not popular, sometimes dye is added to darken the shade and to produce what is called blue-black ink. This name is unfortunately chosen, because it is commonly understood to mean iron gallotannate ink.

This bureau has not spent much time in experimenting with dyes for use in Prussian blue inks, because they are not used by the Government, and are of little commercial importance. One would think first of trying water-soluble nigrosine (C. I. 865; Sch. 700), because it is easily obtained, and is the dye generally used in making ink powders and tablets. Only a limited proportion of nigrosine can be dissolved with the Prussian blue and oxalic acid, without forming a gummy deposit. Three other black dyes, durol black B (C. I. 307; Sch. 265), Columbia fast black FF (C. I. 539; Sch. 436), and direct deep black RW (C. I. 582; Sch. 463) were tried, but with indifferent results. In each case a solution of the dye in distilled water was mixed with a solution of 10 g of Prussian blue and 2 g of oxalic acid in 1 liter of water. When the mixtures stood in corked test tubes for a week, there was no visible settling out of the color, yet they had a curdled appearance when used as writing ink, with a supposedly gold-plated pen.

Those who are sufficiently interested might hunt for a yellow and a red dye that can be mixed with the Prussian blue solution. If the right proportions are hit upon by trial, a practically black solution will be obtained.

4. COLORED DRAWING INKS

In telling about carbon inks it was pointed out that black drawing ink is not a solution, but merely a suspension of carbon in a liquid vehicle. A few of the drawing inks of other colors than black are suspensions of pigments in a liquid. The pigments are usually dye lakes, formed by precipitating a dye on an inert material, such as aluminum hydroxide, barium sulphate, or some other compound. It is filtered from the solution, dried and ground. As the lakes in general are of higher specific gravity than carbon, and are more compact at the outset, the problem of grinding them so fine that they will not settle to the bottom of the liquid is more serious than with carbon. The few pigment inks that have been received by the bureau had a decided tendency to separate. Most colored drawing inks are clear solutions of dyes. They do not have the same degree of hiding power as the inks with comparatively opaque pigments, but their good working qualities outweigh the disadvantage due to their being transparent.

At one time there was a Government specification for colored waterproof drawing ink, which contained formulas for making inks of several colors to serve as standards for fastness to light and water. The waterproofing solution was made by dissolving bleached shellac and borax in water; and certain dyes were to be dissolved in this. Although given what was believed to be an adequate test in the laboratory, at the Panama Canal Zone, and in a bureau where a great deal of drafting is done, two or three of the standard inks developed serious faults in the course of a year. It became necessary to cancel the specification, pending more laboratory work.

The production of a really waterproof ink, which when dry will not be blurred by accidental wetting, nor by cleaning with a damp cloth, is not something that can be done offhand. The ink must contain materials that are soluble in water, but become insoluble if they are allowed to dry on paper or tracing cloth. The usual combination is shellac and borax, together with dyes that seem to have some sort of affinity for the other substances, so that they resist the solvent action of water. By no means all dyes are suitable, for many of them can be leached easily out of the dried shellac-borax film.

The Federal specification was canceled because of faults shown by the standard inks when kept, and the development of a satisfactory formula required a great deal more time than had been anticipated when the work was started. It did not suffice to use a more concentrated solution of borax and shellac, but it was found that if the amount of shellac per liter was increased, the quantity of borax lessened, and ammonia added to give sufficient alkali for dissolving the shellac, the inks could be made extremely resistant to water. This was true if the dyes were such that they were held by the shellac-borax film. Otherwise there was no improvement.

It would take too much space in this circular to give detailed directions for preparing the solution of purified shellac. It must suffice to say that the shellac was purified by removing the insoluble waxy portion, the orpiment, and the varied assortment of impurities that seem to accompany it always. Bleaching was a useless treatment, because the color of the natural lac dye that gives the solution a purplish hue did not affect light yellow, light green, or any other of the dyes tested. The solution of shellac contained 50 g of actual shellac, 3 g of borax, and a sufficient quantity of ammonia in a volume of 1 liter. At the start, 100 ml of strong ammonia water was added, but during the heating necessary to dissolve the shellac, much of the gas escaped, though always enough to hold the shellac in solution remained. The best of the 92 dyes that were tried, and the quantity of each dissolved in 100 ml of the shellac solution are shown in the table.

FIRST CHOICE

Dye	C. I.	Sch.	Gram
Erythrosin, yellowish	772	591	0. 5
Brilliant orange R	78	79	. 6
Chloramine yellow	814	617	. 4
Brilliant milling green B	667	503	1. 2
Wool blue G extra	736	565	. 5
Methyl violet B	680	515	. 5
Benzamine brown 3GO	596	476	. 8

SECOND CHOICE

Dye	C. I.	Sch.	Gram
Crocein scarlet MOO	252	227	0. 5
Benzopurpurine 10B	495	405	. 6
Orange R	161	151	. 4
Metanil yellow	138	134	. 8
Thiazol yellow	813	198	. 8
Malachite green	657	495	. 8
New methylene blue N	927	663	. 4
Crystal Violet	681	516	. 4
Benzo brown G	606	485	. 6

The dyes marked "second choice" made ink that did not keep quite so well as that made with the "first choice" dyes. Most of them were not discarded until after the inks had been kept nearly 13 months.

5. SHOW-CARD INKS

The bureau has not had occasion to test show-card inks, but some of the formulas found in books require the preparation of a solution of shellac and borax, in which dyes are to be dissolved. The inks are therefore not dissimilar to waterproof drawing inks, but the requirements as to their working qualities are not so exacting as in the case of drawing inks. Those who wish to experiment should read the directions for making the shellac-borax solution on page 24.

6. HECTOGRAPH INKS

The hectograph is a simple device for making a moderate number of facsimile copies of a letter or drawing. The original is pressed face downwards upon a special surface, or pad, composed of gelatin and glycerin, or of clay and glycerin. The pad absorbs part of the ink of the original, and can then be used for printing upon other sheets of paper. Though " hectograph " means " hundred writing," no one should expect to make that number of copies.

Hectograph ink must contain a large proportion of a dye that has good color strength. When a specification for the ink was being considered, numerous samples of ink were made and tested in order to find out which dyes gave the greatest number of good copies. The inks were made according to a formula obtained from the Government Printing Office, but using acetone instead of the alcohol called for. The formula, in parts by weight, is:

Acetone	8
Glycerin	20
Acetic acid, 28 per cent coml	10
Water	50
Dextrin	2
Dye	10

If the unit of weight is 1 g, this will make a little more than 90 ml, or about 3 liquid ounces of ink.

The dextrin is first dissolved in the water, which must be heated, but need not be boiled. Care must be taken not to char the dextrin at the start, when it clings to the bottom of the vessel in a sticky mass. It is safest to heat the mixture by setting its container in hot water. When a clear, or nearly clear, solution is obtained, add the other liquids, but not before it has cooled. Acetone is combustible and quite volatile, but the amount in the ink is not dangerous. Ink of the color desired is made by dissolving 10 parts by weight of dye in 90 parts of the stock solution.

The best dye of all is methyl violet B (C. I. 680; Sch. 515). Crystal violet (C. I. 681; Sch. 516) is nearly as good. For red ink, rhodamine B (C. I. 749; Sch. 573) was selected, with fuchsine, or magenta (C. I. 677; Sch. 512) as second choice. The two reds are quite different in shade. Fuchsine is the only exception among all the dyes tested, for because of its slight solubility, only 3.5 parts of it could be dissolved in 90 parts of the solvent. For green and blue, emerald green (C. I. 662; Sch. 499) and Victoria blue B (C. I. 729; Sch. 550) ranked first, with malachite green (C. I. 657; Sch. 495) and soluble blue (C. I. 707; Sch. 539) second.

With a clay-glycerin pad and inks made with these dyes, it was possible to get at least 30 copies in which the strokes of the pen were unbroken, and numerous other copies that were easily legible, though with more or less broken lines. With some of the dyes tested this could not be done.

There is no truly black water-soluble dye. The first hectograph copies made with ink containing nigrosine were not black, and the later copies were of a dingy purplish gray. It is possible to make black ink by mixing dyes; for instance, green, violet, and yellow, in the right proportions, but as the dyes are not absorbed by the hectograph pad or by paper in the same proportions, no black copies could be made.

A bright yellow ink, made with auramine (C. I. 655; Sch. 493), was almost illegible, on account of a curious optical effect. Its brightness was nearly the same as that of the white paper, and the copies looked extremely blurred. When seen through blue glass, they were sharp and distinct.

7. STAMP-PAD INKS

Federal specification TT–I–556, Ink, Stamp-pad, gives a formula for a standard in several colors. The vehicle, or liquid portion, consists of 55 parts by weight of glycerin, and 45 parts of water. In this is dissolved 5 parts of one of the dyes mentioned further on.

At ordinary temperatures glycerin practically does not evaporate, and it is hygroscopic, or attracts moisture from humid air. For these reasons it prevents the ink from drying on the pad, even in winter, when the air in most heated buildings is of desert aridity. In summer, when the air is usually of high humidity, the ink tends to take up water. In neither case is the change in the ink sufficiently great to affect its use.

The dyes required by the specification are fuchsine (magenta; C. I. 677; Sch. 512), light green SF (C. I. 670; Sch. 505), soluble blue (C. I. 707; Sch. 539), acid violet (C. I. 698; Sch. 530), and nigrosine (C. I. 865; Sch. 700). A great variety of other dyes can be used, if desired.

8. RECORDING INKS

There are numerous types of instruments for making continuous records of temperature, barometric and steam pressures, electric voltage, etc. The record consists of a line or a series of dots on a circular card or a long sheet of paper. The instrument may have to run a long time without attention, so the ink must not dry on the pen; and if it is located outdoors, the ink must not freeze. For many years the United States Weather Bureau has used recording ink made exactly like stamp-pad ink, but with much less dye. The mixture of glycerin and water is a good "antifreeze," yet there are parts of the country where it would be solid in winter. To overcome this difficulty, enough alcohol to keep the ink fluid at any outdoor temperature is mixed with the ink. For indoor use, a mixture of 1 volume of glycerin and 3 volumes of water has been found satisfactory for recording ink.

The stamp-pad ink of the specification contains 50 g of dye in 1 liter of solvent, but only one-fifth as much dye suffices for a liter of recording ink. Dyes that have been found to work well in recording inks are: Crocein scarlet MOO (C. I. 252; Sch. 227), fast crimson (C. I. 31; Sch. 42), orange TA (C. I. 374; Sch. 311), crocein orange (C. I. 26; Sch. 37), brilliant yellow (C. I. 364; Sch. 303), emerald green (C. I. 662; Sch. 499), soluble blue (C. I. 707; Sch. 539), methylene blue (C. I. 922; Sch. 659), methyl violet B (C. I. 680; Sch. 515), Bismarck brown R (C. I. 332; Sch. 284), and nigrosine (C. I. 865; Sch. 700).

9. INDELIBLE MARKING INKS FOR FABRICS

Federal specification TT–I–542, Ink, Indelible, Marking (for) Fabrics, gives no formula for a standard ink. Of the numerous formulas found in books, nearly all are either silver inks or aniline black inks.

Although silver ink looks black in the bottle and makes dark marks on white fabrics, its full color must be developed by heat, or by exposure to bright light for a time. This causes a precipitation of metallic silver in and on the fibers. The metal is so finely divided that it looks black. An old formula, slightly modified, is here given. Dissolve 5 parts of silver nitrate in 5 parts by weight of water, and add ammonia water in small amounts until the precipitate that first forms is dissolved. In separate vessels dissolve 3 parts of anhydrous sodium carbonate in 15 parts of water, and 5 parts of gum arabic in 10 parts of warm water. Pour the three solutions together and warm gently until the mixture starts to darken. If 1 part equals 1 g, the formula will make about 35 ml, or a little more than 1 liquid ounce of ink.

"Household" ammonia is generally not suitable, because it contains other substances than ammonia gas and water. Pure ammonia water can be obtained from a druggist, and he will be able to supply sodium carbonate monohydrate, if not the anhydrous form. The monohydrate contains 14.5 per cent of water of crystallization, so 3.5 parts of it must be used, as the equivalent of 3 parts of the anhydrous salt.

Metal vessels should not be used for preparing the solution of silver nitrate, nor for the finished ink. It is simplest and safest to dissolve the nitrate in the bottle in which the ink is to be kept. The salt is easily soluble, and it is not necessary to use heat to dissolve it. The other two solutions can be poured into the bottle, and if they are somewhat warm it may require no further heating to darken the mixture. This final step in the preparation is not necessary, but is convenient because it is easier to use a dark ink than a pale one. If the freshly mixed ink does not darken, set the bottle cautiously in a vessel of warm water.

Because of the action of steel upon silver ink, a gold pen is the best to use or, if it can be had, a quill pen. Lacking these, use a new steel pen, or a gold-plated one. Then, when the marks are dry, press them with a hot flatiron, or place them in full sunlight, to develop the black color. The marks will then be very resistant to washing, unless the laundry uses too strong a solution of chlorine as the "bleach," in which case the silver will soon be converted into silver chloride. This compound is not very soluble in water, yet sufficiently so as to be washed out.

There are formulas for silver marking inks colored with dyes, but these seem to have no advantage over the black ink.

Aniline black is extremely fast to light and washing, so it is suitable for marking ink. The dye is not soluble in any liquid that could be used on clothing, but it can be formed readily on and in the fibers of cloth. The bureau has not made inks of this kind, but the formula about to be given is nearly the same as that for blackening wooden table tops, which has been used and found satisfactory. It consists of two solutions which are mixed immediately before use, if clothing is to be marked. For staining wood, the solutions are applied alternately until the desired depth of color is produced.

The formula is:

Solution A—
    Copper (cupric) chloride_____ 85
    Sodium chlorate_____ 106
    Ammonium chloride_____ 53
    Water_____ 600
Solution B—
    Gum arabic_____ 67
    Aniline hydrochloride_____ 200
    Water_____ 335

For use, mix 1 volume of A and 4 volumes of B. The marks should not be exposed to sunlight nor ironed to develop their color.

Sodium chlorate is quite different from common salt, sodium chloride. The chlorate when mixed with combustible substances is extremely dangerous, because friction or shock may cause the mixture to explode with great violence. Potassium chlorate, which is also dangerous, is easier to obtain than the sodium salt. If used, 122 parts of it is equivalent to 106 parts of the sodium salt.

Aniline black is not produced the instant the two solutions are mixed, but is formed gradually, so the mixture has a chance to penetrate into the fibers and form the dye there, instead of in a layer on the surface. This may be considered as an argument in favor of taking the extra trouble to prepare the 2-solution ink; but there are excellent 1-solution aniline black inks on the market.

10. SYMPATHETIC INKS

Persons who indulge in secret writing for legitimate or nefarious reasons must have invisible or sympathetic ink. In their chapter on this subject, Mitchell and Hepworth [6] say that both Ovid (43 B. C. to A. D. 17) and Pliny (no doubt the Elder, A. D. 23 to 79) tell about sympathetic inks. They knew of the use of milk and plant juices for this purpose. When heated moderately, the writing turns brown before the paper or parchment begins to scorch, and thus the message becomes readable. These are but two examples of secret inks that are made visible by heating. Some of the materials that are used char more easily than paper, but others act upon the paper in such a way that the writing turns brown, or even jet-black.

Without going outside of the home, one can get several materials for making sympathetic ink that is developed by heat. Any of the following substances can be used, though it must be confessed that some of them are very poor indeed: Alum, soda (either baking or washing), borax, flour boiled with plenty of water, corn sirup thinned with water, soap solution, diluted mucilage, milk, lemon juice, saliva, or storage battery acid diluted with 10 or 12 volumes of water.

The salts of several metals have long been favorite materials for sympathetic ink. These salts are not all colorless in the solid form or in strong solution, but solutions so dilute as to make invisible marks on paper can be developed by suitable means. Among

[6] Mitchell, C. A., and Hepworth, T. C., Inks, Their Composition and Manufacture, 3d ed. Chas. Griffin & Co. (Ltd.), London, 1924.

these salts are lead acetate, ferric sulphate, mercuric chloride (corrosive sublimate, dangerous to handle and very poisonous), copper sulphate, cobalt chloride, and nickel chloride. In addition to being turned brown or black by the vapors of ammonium sulphide, writing with any of these salts can be developed by heat, and still other means can be employed with some of them. For instance, if a ferric salt is used, tannic acid solution will turn the writing black, or potassium ferrocyanide will form Prussian blue.

Of the salts just mentioned, cobalt chloride is in some respects the most interesting. When a water solution of the salt evaporates to dryness, the chloride appears in crystals which are red, though not intensely so. If the solution used as sympathetic ink is very dilute, the thin layer of crystals left in the paper when the writing dries will not be perceptible. If the writing is kept for some time in quite dry air, or is warmed moderately, the cobalt chloride loses most of its " water of crystallization " and then is so intensely blue that the writing is visible. Exposure to moist air, as by breathing upon it, makes the writing vanish, because the blue salt regains its water of crystallization and turns red. These changes back and forth can be repeated many times, but if once the secret writing should be heated a little too strongly, the paper will char and the writing will be permanently black.

As a means of developing secret writing with a variety of substances, iodine is interesting. It is preferably used as the vapor from the solid element, though the tincture diluted with water can be employed. If a thin solution of starch was used for the writing, iodine will turn it blue. The color disappears after a time and more quickly by gentle warming. Writing with a solution of soap becomes yellow or brown because the fatty acids in the soap absorb iodine vapor more readily than the paper does. This color soon disappears, for iodine is so volatile. Copper sulphate and lead acetate are also colored temporarily, while marks made with mercuric chloride show as white on a background of yellow paper. Finally, if the writing was done with distilled water, iodine vapor will color the letters a little more strongly than the background. The water disturbs the sizing at the surface of the paper, and thus allows the iodine vapor to be absorbed more readily there than elsewhere.

In the examination of a document suspected of containing secret writing, the first step would naturally be to heat the paper moderately. This may bring out nothing, but it is not apt to destroy the latent evidence. Exposure to iodine vapor or to the fumes of ammonium sulphide might be next in order, though the fumes from ammonia water could be used. If heat and the various vapors fail, chemical solutions must be tried on selected small parts of the document. To treat the whole sheet of paper with a reagent that brought no visible result might destroy all chance of developing the writing.

11. INKS FOR SPECIAL SURFACES

All the inks so far discussed in this circular have been intended for use on paper or similar materials, and they are not well adapted to writing on impervious or oily surfaces, such as glass, porcelain, celluloid, metals, or painted articles. It is true that by going over the lines repeatedly, it is possible to write after a fashion on some of these materials with iron gallotannate and drawing inks, but the behavior of the ink shows that it is not suitable for the surface. When the pen is lifted it leaves a surplus of ink at the end of the line which is elsewhere pale; and the dried line is only a loosely attached crust that can be rubbed off easily.

(a) INKS FOR CELLULOID

Trial of two published formulas for inks to be used on celluloid showed that they would make permanent marks, because they contained acetone, a solvent for celluloid. The trouble was that they spread excessively over the surface, so that with a fine pointed pen the narrowest line that could be drawn was about one-eighth of an inch wide.

Some years ago it was found that a commercial solution of bitumen in coal-tar naphtha could, though with a little trouble, be used for writing on celluloid. It was necessary to dip the pen and, within a second, touch it to the celluloid, or the point would become dry and the ink refuse to flow. The slightly raised letters were quite resistant to rubbing, though they could be erased by means of absorbent cotton moistened with benzol.

Success with this solution led to experimenting with solutions of asphalt in other solvents, and it was found that amyl acetate (banana oil) gave the best results. However, if the solution were too concentrated, in the attempt to make black ink, it could not be used. A weaker solution made rather sharp lines of a dark brown color.

(b) INKS FOR GLASS AND PORCELAIN

Some of the inks recommended in books for writing on glass contain sodium silicate solution, or water glass, mixed with pigments that resist the action of the alkali in the silicate. Water glass should not be used unless the marks are expected to last as long as the glass, for when the solution dries it forms such a strong bond with the glass that it can not be removed completely without grinding it off.

Solutions of resins and dyes in volatile, combustible solvents have also been suggested, but as the marks can be removed readily by means of solvents, one might as well use a safer ink that works quite well on porcelain, glass, and incidentally, on aluminum. The solvent is made by heating, but not boiling, a mixture of 4 parts by weight of dry shellac, 1 part of borax, and 150 parts of water. The solution must be filtered, preferably after it has cooled, to remove the insoluble waxy portion, the orpiment that settles to the bottom, and the miscellaneous impurities that shellac always seems to contain. The purplish color of the solution will not interfere noticeably with the color of the dyes that are dissolved to make the ink. (This solution is not satisfactory for colored waterproof drawing ink.) From one-half to 1 g of dye will usually suffice for making 100 ml of ink.

The following are suggested in addition to those mentioned in connection with colored waterproof drawing ink:

	C. I.	Sch.
Naphthol yellow	10	7
Tartrazine (orange in this ink)	640	23
Diamine sky blue FF	518	424
Naphthol blue-black S (greenish blue)	246	217
Benzo cyanine R (Verging on Violet)	405	336
Durol black B (blue-black)	307	265
Nigrosine (purplish or bluish black)	865	700

It should not be thought that this ink can not be washed from glass. To get such a degree of permanence water-glass ink or actual etching must be resorted to.

(c) ETCHING INK FOR GLASS

Bureau of Standards Letter Circular No. 150, Dry Etching of Glass, gives detailed directions for using the method by which certified burettes and other glass measuring apparatus are marked. Hydrofluoric acid, which is used in one method, is a very dangerous substance in the hands of persons who are ignorant of its properties or are careless. It produces severe sores on the skin, which are slow to heal. In addition, the fumes are harmful to breathe, would probably seriously injure the eyes, and will corrode metals. Spectacle lenses are quickly and permanently clouded if the air contains any considerable quantity of the fumes.

(d) INK FOR ZINC GARDEN LABELS

Ink for garden labels made of sheet zinc is of interest to many gardeners. Perhaps the most durable marks are made with a solution containing copper, because as the zinc slowly weathers away the traces of copper in the marks keep up an electrolytic action, so that the writing remains as slightly sunken black lines on the zinc. Labels written with this sort of ink will not be obliterated by exposure to the weather for five or six years in the climate of Washington. If the writing becomes obscured by the products of the corrosion of the zinc, all that is necessary to restore their legibility is to rub them with a finger to remove the coating.

A good copper ink is made by dissolving 1 part each by weight of copper acetate and ammonium chloride in 15 parts of water. With this is mixed enough lampblack to give a good color. Without the lampblack the other ingredients do not make a clear solution, because a comparatively insoluble double salt is formed as a light blue deposit on the bottom of the bottle. The solution above it is of a much darker shade of blue. Possibly when the lampblack is present, the double salt forms on the particles of carbon and is thus kept from caking on the bottom of the bottle. There is no other apparent reason for using lampblack, because the clear solution will make black marks instantly, though they are perhaps not as intensely colored as those made with the next ink to be described.

(e) INK FOR BRASS

Instrument makers sometimes give brass a mat black finish by dipping the perfectly clean metal into a solution made by dissolving copper carbonate in ammonia water. This suggested making an ink by dissolving 1 part of copper acetate in 15 parts of water, and adding to this enough ammonia water to dissolve the blue precipitate that is first formed. A similar solution made with copper sulphate does not make as black marks. Neither solution can be used on copper, because the blackening is caused by a chemical reaction between the copper in solution and the zinc in the brass.

(f) INKS FOR OTHER METALS

A dilute solution of silver nitrate makes black marks on copper, brass, tin, and some other metals. The marks are still better if ammonia water is added to the solution of silver nitrate in water. The clear solution makes beautifully sharp, black lines. It has the disadvantage that the silver salt formed by the action of ammonia has been known to explode with great violence if the solution is allowed to evaporate. The crusts that sometimes form around the cork are the explosive compound. If any such deposit is noticed, it should be rinsed off. To try to wipe it off may cause it to explode.

For writing on aluminum the solution of shellac, borax, and dye that was recommended for glass is quite satisfactory.

(g) TIME-CARD INK

On letter boxes in smaller towns the hours at which mail is collected are written on white lacquered cards. The specification under which the Post Office Department buys the time card or letter-box ink requires that the ink supplied on contract shall be as resistant to weather as one made as follows: Mix 25 parts of shellac varnish (4 pounds to the gallon), 10 parts of denatured alcohol, and 15 parts of technical cresol. In this dissolve 5 parts of nigrosine base (C. I. 864; Sch. 698). This ink can be used for writing on a variety of surfaces.

IV. TESTING INKS

An essential part of each Federal specification for an ink is the section in which the methods of testing are described. If the specification includes a formula for a standard ink, both the sample and the standard are subjected to the same tests, and as far as possible on the same sheets of paper. The chief reason for having standard ink is to use it as a basis for comparison in the testing, because similarity or dissimilarity in behavior becomes apparent at once. Even if the sample is a red ink, or a stamp-pad ink, it need not match the standard in color, unless this has been agreed upon by buyer and seller. The sample must, however, be at least as satisfactory in working qualities as the standard, and as fast to light.

1. IRON GALLOTANNATE INK

Both kinds of iron gallotannate ink can be discussed together, because exactly the same test methods are employed. The tests follow naturally from the definition of ink given by Schluttig and Neumann, which is quoted on an earlier page of this circular. The definition clearly tells what properties an iron gallotannate ink should have, and the tests are intended to find out whether the sample has these properties. In the fifth chapter of their book, their testing procedure is given, and also detailed explanations of each test. Some of their tests are not required by the Federal specifications, but an important one that they omit (the corrosion test) is included.

Because freshly made, well-settled ink should be clear, the first step in the examination of a sample is to allow the ink to remain undisturbed for 24 hours. If the sample is in several small bottles, which together contain the pint of ink called for, the contents of the bottles are poured together into a large one of a suitable size. If the sample is concentrated ink, it is diluted with the requisite volume of water and thoroughly mixed. Powders and tablets are dissolved in water. In any case, after 24 hours the bottle is slowly tilted, while being held up against the light, and is observed closely to see whether any sediment has settled to the bottom. There should be at most only traces of sediment.

If the ink passes this inspection, the test for keeping quality is started next, because it takes two weeks to complete. In two similar clear glass vessels, for instance, crystallizing dishes 35 mm (1⅜ inches) deep, and 48 mm (1⅞ inches) in diameter, are placed 25 ml portions of the sample and the standard. The dishes, loosely covered, are kept where they will be in diffused daylight, but never in direct sunlight, for two weeks. The sample should show no more surface skin than the standard, nor more deposit on the walls and bottom of the container.

The iron content of the sample is determined in a 10 ml portion, by any suitable analytical procedure. The amount of iron in 100 ml of copying and record ink should not be less than 0.58 g, nor more than 0.70 g. For writing ink the limits are 0.29 and 0.35 g.

Streaks are made side by side on white bond paper with the sample and the standard. The sheet of paper, about 8 by 10 inches, is pinned to a board or clamped to a pane of glass, and held at an inclination of about 45°. Measured portions, of 0.6 ml each, of ink are allowed to flow down across the sheet. The ink is measured in a pipette made of glass tubing of about 3.5 mm (0.138 inch) bore and 250 mm (about 10 inches) long. A mark etched or scratched 62 mm ($2\frac{7}{16}$ inches) from one end of the tube indicates the required volume of ink. The ends of the tube can be fire-polished, but should not be constricted. Ink is drawn up to the mark, and kept from flowing out by placing a finger tip against the upper end of the tube. While holding the tube vertically, place its lower end against the upper edge of the inclined sheet of paper, remove the finger, and allow the ink to flow out all at once, and down across the paper. Make one or two more streaks with the sample ink and then, with another pipette, make streaks with the standard, close beside those of the sample. The sheet of paper is left in position until the streaks are dry, and is then put where it will be in diffused daylight, but not in direct sunlight.

It is not necessary, according to the specification, that the freshly made streaks of the sample and the standard shall be of exactly the same color, though they should be equally uniform in color from top to bottom. They should be of about the same general shape and the same width, because these features indicate that the inks are about equally fluid—an exact measurement would be a waste of time. The streaks made with the sample should show no more evidence of striking through the paper than do those of the standard.

After being kept in diffused daylight for a week, the streaks of the sample should be as intensely black (in reality blue-black) as those of the standard. The sheet of paper is cut across the streaks into strips about 1 inch wide. A strip is soaked in water, and one in 50 per cent denatured alcohol, for 24 hours. A third strip is exposed to direct sunlight for 96 hours, or at a distance of about 25 cm (10 inches) from an arc or ultra-violet light for 48 hours. In these three tests the sample should retain its color as well as the standard.

A final test for stability of color is to soak a strip in bleaching powder solution containing $N/200$ available chlorine. The two sets of streaks are compared after 15 minutes, 1 hour, and 24 hours in this solution at room temperature.

Because of the temptation to increase the amount of free mineral acid in the ink, in order to delay the deposition of sediment, a test of the corrosive action of the ink upon steel pens is made. This test is of no interest to the millions of users of fountain pens, because gold is not attacked by the acids in ink. However, if they are writing something that must be kept as a record for a long time, the acid in the ink, especially if sulphuric, may have been used in such an amount as to damage the paper eventually. The other millions of writers who use steel pens must be looked out for. The Post Office Department, alone, asked for bids on 5,212,800 steel pens for the fiscal year beginning July 1, 1932.

The amount of corrosion of steel pens is something more than a measure of the free acid in the ink, because other substances in the ink may promote the corrosion. Common salt is a familiar example of the many substances that cause iron to rust rapidly in the presence of moisture and air. Other substances retard corrosion, notably the soluble chromates.

Take two new steel pens, from the same box, for each sample of ink, including the standard. Rinse the pens with alcohol, then with ether, and dry them in an air oven at 105° C. (221° F.). Weigh each pair together to the nearest milligram. As the treatment with the solvents is intended to remove the oily film that is on new pens, it is advisable to handle them with forceps. Immerse each pair of pens in 25 ml of ink, contained in a small beaker or flask, taking care not to have them "nested" together, but separated from one another. After 48 hours remove the pens, clean them with water, and by rubbing with a cloth to remove the tightly clinging deposit, rinse with alcohol, and dry at 105° C. Again weigh, and if the loss in weight of the pair of pens in the sample is greater than the loss of those in the standard, the test should be repeated with both inks and new pens.

All of the metal dissolved from the pens is taken from their surfaces, and the amount removed therefore depends to some extent upon the area of the metal surface in contact with the ink. This would not be true if corrosion stopped when all the free acid was neutralized by iron. As already pointed out, part of the corrosion may be due to other substances in the ink which react with the iron in the presence of moisture and air. The area of exposed metal has no direct connection with the weight of the pens. In an example to be cited a little further on, a pair of pens used in testing an ink weighed 1.010 g, while another pair from the same box weighed 1.068 g, an increase of about 6 per cent. Assuming that the pens were stamped out with the same die, the areas of their broad surfaces must have been the same within quite narrow limits. The areas of the edges were 6 per cent greater in one case than in the other, if the heavier pair were thicker than the others, which seems to be a fair assumption. An increase of 6 per cent in the comparatively small area of the edges would affect the total area very much less than that percentage.

Because the weights of the pens may vary a few per cent, while their areas are almost the same, it is inadvisable to express corrosion losses as percentages of the original weights of the pens. In fact, it may happen that expressing the losses as percentages will reverse the actual results of a comparative test. In a test of several samples of ink the weights of pairs of pens ranged from 1,010 to 1.068 g; and the losses in two of the inks were 0.057 and 0.059 g. If the greater loss had happened to the heavier pair of pens, it would be the smaller if calculated as a percentage.

The amount of corrosion depends also upon the composition of the metal and upon the treatment to which it has been subjected while making the pens. These are further reasons for taking all the pens for a given test from the same box, it being assumed and hoped that they will be more nearly alike in all respects than peas from the same pod.[6]

As a rule, all parts of the pens seem to be equally attacked by the ink, but occasionally a strikingly different type of corrosion is met with. The attack is chiefly at the edges, including those of the open slot and the slits. Pens have been seen with their central slits opened to a width of nearly a millimeter (one twenty-fifth of an inch). This effect, which seems not to have been mentioned elsewhere in print, is due to the ink, because both pens are always affected in the same way and to the same extent in a given ink. A possible reason is that the ink contains an acid that is barely able to dissolve iron, so its action is limited to the part of the metal that has undergone the most severe mechanical treatment. This, of course, is where it has been cut.

Other ways of determining the corrosiveness of inks have been suggested.[7] The most unexpected is that of Schluttig and Neumann, who said:

We can be sure that an ink which darkens as quickly and intensely as the type [standard] can not contain relatively too much free acid—relatively only, for according to the quantity of iron salts the permissible degree of acidity will be greater or less, naturally only within narrow limits.

2. RED INK

Red ink is much simpler to test than blue-black ink. It is examined for sediment in the same way, and is judged more strictly from that standpoint, because a manufacturer has no excuse for failing to make a clear solution of a dye. Streaks are made on paper, and are subjected to the light-fading test, but for only half as long as blue-black ink.

3. STAMP-PAD INK

Stamp-pad inks dry almost entirely by being absorbed by the paper on which the impressions are made. That they can not dry to any great extent by evaporation is evident from the fact that they contain a large proportion of glycerin, which tends to absorb moisture from the air and is practically nonvolatile at ordinary temperatures. For these reasons it is used to keep the ink from drying on the pad. For testing, small stamp pads are made by cutting disks of white felt about 6 mm (one-fourth inch) thick and 38 mm (1½ inches) in diameter. These fit snugly inside of rings cut from brass tubing. This kind of pad gives just as good results as the more complicated arrangement described in the specification.

---

[6] Mitchell and Hepworth, on p. 162 of the third (1924) edition of their book, say that this corrosion test is a " somewhat crude one," and then add : " This method, which the writer [presumably Mitchell] devised some years ago and published in the first edition of this book, has recently been included in the United States specifications without acknowledgment." On p. 123 of the first (1904) edition is found the following description and discussion of the test: "Another practical test is to immerse a steel pen in the ink for a given period, and to determine the loss in weight. Thus in the case of the ink referred to above we found that a pen had lost 5.18 per cent of its weight after being kept in 10 c c of the ink for a month, whilst the ink itself had become nearly solid." The germ of the method in the specification may have come from the book, but at least as long ago as 1907, it had been developed to its present improved form, which differs in several details from Mitchell and Hepworth's procedure.

[7] F. F. Rupert, Ind. & Eng. Chem., vol. 15, pp. 489-493, 1923. Mitchell and Hepworth, p. 163 of their 1924 edition; see also Analyst, vol. 46, p. 131, 1921. Schluttig and Neumann. pp. 77-78.

The pads are placed upon a pane of glass, partly for cleanliness, but also to prevent the loss of ink that would occur if they rested upon paper or unpainted wood. Equal volumes of the sample and of the standard ink of the same color are placed upon separate pads. Impressions are made upon the same sheet of white bond paper, with a clean rubber stamp. The impressions made with the sample should dry as rapidly as those made with the standard, and should be as sharp and as intensely colored. Impressions made with each ink are half covered with black paper, and exposed to direct sunlight for 48 hours, or at a distance of about 10 inches from an arc or ultra-violet light for 24 hours. The sample should fade no more than the standard. The inked pads are allowed to stand exposed to the air for 10 days. At the end of that time the sample should show no more evidence of absorption of excessive moisture from the air, or of drying and caking on the pad, than does the standard.

4. INDELIBLE MARKING INK

There is no formula for a standard ink in the Federal specification for indelible marking ink for fabrics. The tests are intended to represent severe treatment in a laundry, and also to find out whether the ink weakens, or " tenders " fabrics. The first step is to prepare inked strips of cloth for the tendering test. Strips 4 inches wide and 36 inches long are cut in both the warp and the filling directions from suitable closely woven cotton or woolen cloth. Each strip is cut into test pieces 6 inches long. Across half of the test pieces representing each direction of the weave, the ink is applied in a band about 1 inch wide. The remaining test pieces are left uninked. After 10 days the breaking strengths of the inked and uninked pieces are determined. The breaking strength of the inked fabric must be not less than 90 per cent of the breaking strength of the uninked fabric.

On other pieces of suitable fabric, marks are made with the ink strictly according to the manufacturer's directions. Some of the marked pieces are kept for two weeks, and are then examined for any discoloration beyond the limits of the actual marks.

Two or three of the marked pieces are put through a series of washing tests in solutions prepared as follows:

Soap solution.—The soap solution contains 7 g of white floating soap and 7 g of modified soda (58 per cent of sodium carbonate and 42 per cent of sodium bicarbonate) in 1 liter of distilled water.

Oxalic acid.—The oxalic acid solution is made by dissolving 6 g of crystallized oxalic acid in 1 liter of commercial 28 per cent acetic acid.

Sodium bisulphite.—The solution contains 5 g of sodium bisulphite and 72 ml of hydrochloric acid of specific gravity 1.11, with water to make 1 liter.

Bleaching solution.—A stock solution containing 1.4 per cent of available chlorine is prepared. For use, 100 ml is diluted with 1,300 ml of water.

All the washing tests are made with the solutions at 65° to 71° C. (149° to 160° F.). The marked pieces are immersed in the soap solution for 15 minutes, then rinsed 5 times in distilled water and

Metric weights and measures of capacity are so simple in principle and in their interrelations that they are used in most of the formulas in this circular. As many readers will prefer ordinary weights and measures, some conversion factors and other pertinent information will be given.

The unit of weight in the metric system is the kilogram, or 1,000 g; and the unit of capacity is the liter, or 1,000 ml, which is the volume occupied by 1 kg of pure water at the temperature of its greatest density, and under the pressure of a normal atmosphere. According to the original intent, the kilogram was to have been the mass of 1,000 cubic centimeters (cm^3) of water at its greatest density, and a liter would then have been 1,000 cm^3. This relation was not realized on account of experimental difficulties, and it is now known that instead of being 1,000.000 cm^3, the actual volume of the liter is 1,000.027 cm^3. The difference, equal to the volume of a small drop of water, is negligible in most chemical work. Many laboratories still cling to the name cubic centimeter for the volume called milliliter throughout this circular.

Water expands when heated, and thus its density, or weight per unit volume, decreases. The change is small for ordinary temperatures, and for practical purposes not requiring great accuracy, the weight of both 1 liter and 1,000 cm^3 can be called 1 kg; and 1 ml and 1 cm^3 will weigh 1 g. So if a formula calls for 50 g of water, it will be sufficiently accurate to measure it.

If it is desired to measure other liquids than water, instead of weighing them, their specific gravities must be considered. For intance, if 35.4 g of concentrated sulphuric acid is required, that figure must be divided by 1.84, the specific gravity of the acid. The volume is 35.4/1.84, or 19.2 ml. The rule is the same for liquids lighter than water. Thus, 80 g of acetone, of specific gravity 0.79, will be equal to 80/0.79, or 101.3 ml.

Liquids pack perfectly without air spaces, so it is safe to measure, instead of weighing, them. This is not true of solids, because the weight that a given measure will hold depends upon the size and shape of the particles, and how they are arranged. The specific gravity of the solid is not always the most important factor, as an illustration will show. The oxide of lead known as litharge has a specific gravity that varies somewhat according to the method of preparation, but it is usually a little above 9. Some determinations were made of the weight of a sample of unusually fine-grained litharge that could be packed into the space of 1 cubic inch (16.39 cm³). Taking the specific gravity as 9, a cubic inch of litharge in one solid piece would weigh 9×16.39, or 147.5 g. The weight of the fine powder that could be packed into the cubic inch measure was about 35 g, or 112.5 g less than the solid piece. In other words, the air between the particles of litharge occupied 112.5/147.5, or about 76 per cent, of the cubic inch. The apparent specific gravity of the powder was 35/16.39, or 2.14.

The conversion factors about to be given are based upon accurate comparisons of the kilogram (kg) and the avoirdupois and apothe-caries' pounds, and of the liter and the United States gallon. The avoirdupois pound is 16 ounces, and the apothecaries' pound, 12 ounces.

Metric and avoirdupois weights

1 kg=2.2046 pounds=35.274 ounces.
1 g=0.001 kg=0.0353 ounce=15.43 grains.
1 pound=0.45359 kg=453.59 g.
1 ounce=28.35 g.

Metric and apothecaries' weights

1 kg=2.6792 pounds apothecaries'=32.151 ounces apothecaries'.
1 g=0.0322 ounce apothecaries'=15.43 grains.
1 pound apothecaries'=0.37324 kg=373.24 g.
1 ounce apothecaries'=31.103 g.

Metric and United States measures

1 liter=0.2642 gallon=1.0567 quarts=33.81 liquid ounces.
1 ml=0.001 liter=0.0338 liquid ounce.
1 gallon=128 liquid ounces=3.7853 liters.
1 quart=32 liquid ounces=0.9463 liter=946.3 ml.
1 liquid ounce=29.57 ml.
1 liter of water weighs 1 kg, or 1,000 g.
1 gallon of water weighs 8.33 pounds.
1 liquid ounce of water weighs 1.04 ounces avoirdupois.

If the weight of a gallon of water is calculated from the factors, 2.2046×3.7853, the result is 8.345 pounds. The discrepancy between this and 8.33 pounds is caused by the different temperatures at which the liter and the gallon measures are filled with water.

The British, or imperial, gallon holds 10 pounds of water, and is therefore 1.200 times the United States gallon, and the latter is 0.833 times the imperial gallon. So, for instance, 1 liter is 1.0567×0.833 imperial quarts, and 1 imperial quart is 0.9463×1.200 liters.

The figures to the right of the decimal points can not be ignored in making calculations, but when it comes to actually weighing or measuring the materials for a liter of ink, even the first decimal is usually of little significance. At the same time it must be remembered that kitchen scales are not suitable for weighing grams, nor a battered quart cup for measuring liters. One must not be too haphazard when making the standard ink of a specification.

The use of the conversion factors is seen from the following example. Suppose it is desired to make 125 gallons of the standard red ink which contains 5.5 grams of dye per liter. One gram equals 0.03527 ounce, and 1 liter equals 1.0567 quarts; so 1 gram per liter equals 0.03527/1.0567, or 0.03338 ounce per quart. Then 5.5 grams per liter is 5.5×0.03338, or 0.1835 ounce per quart. This last figure multiplied by 4 to get ounce per gallon, and then multiplied by 125 to get the weight of dye for 125 gallons, gives 91.78 ounces, or 5.74 pounds of dye.

2. EQUIPMENT FOR MAKING INK

The manufacturer of inks will have proper equipment, but whoever makes small batches at home must put up with makeshifts, unless he has some chemical glassware for preparing the solutions, a measuring cylinder or two, and moderately sensitive scales with small weights. Dealers in photographic supplies sell cheap scales for those who make their own developing solutions. Usually the weights are apothecaries'.

If ordinary bottles must be used and the solutions have to be heated, there are two safe ways of going about it. One is to set the bottle in a saucepan containing water, and then to pour in hot water slowly, and not against the side of the bottle. Another way is to set the bottle in water as before, but to put under it a piece of wire netting or a flat spiral of heavy wire to prevent the bottle from touching the bottom. Then it can be heated over a gas flame or on a stove. The idea in either case is not to heat the outside of the bottle too quickly while its contents are cold. The materials will dissolve more rapidly if the bottle is swirled or shaken frequently so as to stir up the relatively concentrated solution that settles to the bottom.

So far as possible, avoid the use of metal vessels for making ink. Iron, especially, should be avoided because it is acted upon so easily by acids, and may also cause discoloration of some dyes.

3. DYES FOR MAKING INK

Not all dyes are equally suitable for making inks, and in the formulas in the body of this circular various dyes have been recommended for one or another of the inks. In each case the name of the dye is followed by certain letters and numbers in parenthesis; as, "Soluble blue (C. I. 707; Sch. 539)." An explanation of these symbols is in order.

There are many more dye names than there are different kinds of dyes, because manufacturers like to use names of their own choice for their products. As a rule, the more widely a dye is used, the more apt it is to have a great many names. At least a dozen names have been given to the familiar dye, Bismarck brown. This practice would of itself cause confusion, and it is further complicated by the fact that nearly or quite the same name may be given to dyes that differ in chemical composition and structure; thus, there are several fast reds and soluble blues. It is true that some attempt is made to distinguish between different dyes that have the same name by adding letters, or letters and numerals, after the names. This plan would be better if there were not so many letters in the alphabet. To give an example, the same dye has been called erythrosine, or erythrosine followed by D, B, J, JNV, or W, to say nothing of seven other quite different names that have been given it. Some names of this kind are in the list given further on.

To do away with this sort of confusion, two important tabulations of dye names have been published. The first is Gustav Schultz's Farbstofftabellen (Dye Tables), and the second is the Colour Index of the British Society of Dyers and Colourists. In each book the dyes are arranged in order according to their chemical nature, and they are numbered serially. No. 707 in the Colour Index refers to a particular dye, one of the soluble blues, and to nothing else. The same dye is No. 539 in Schultz's book. It will now be evident to the reader that such a symbol as (C. I. 707: Sch. 539) is the most certain way of telling the seller just what dye type is wanted. He will probably have both books, but if he has only one, the advantage of giving both numbers is obvious.

The Year Book of the American Association of Textile Chemists and Colorists gives the names by which the various types of dyes made in this country are known by the manufacturers. Usually there are several names given to each type. With one or two exceptions, the names in the following list are those given in the Year Book, in parenthesis immediately under the respective Colour Index numbers. A few alternative names are also given.

RED DYES

	C. I. No.	Schultz No.
Fast crimson	31	42
Azorubine (nacarat B)	179	163
Crocein scarlet M O O	252	227
Benzopurpurine 10B	495	405
Fuchsine (magenta)	677	512
Rhodamine B	749	573
Erythrosin, yellowish	772	591

ORANGE DYES

	C. I. No.	Schultz No.
Crocein orange	26	37
Brilliant crocein R	78	79
Orange R	161	151
Orange T A	374	311

YELLOW DYES

	C. I. No.	Schultz No.
Naphthol yellow	10	7
Metanil yellow	138	134
Brilliant yellow	364	303
Tartrazine	640	23
Auramine	655	493
Thiazol yellow	813	198
Chloramine yellow	814	617

GREEN DYES

Malachite green (Victoria green)	657	495
Emerald green	662	499
Guinea green B	666	502
Brilliant milling green B	667	503
Light green SF	670	505

BLUE DYES

Naphthol blue-black S (agalma black 10B)	246	217
Benzo cyanine R	405	336
Benzo blue 2B	406	337
Diamine sky blue FF (direct sky blue 6B)	518	424
Benzo sky blue (direct sky blue)	520	426
Soluble blue (Bavarian blue DSF)	705	537
Soluble blue	707	539
Victoria blue	729	559
Wool blue G extra	736	565
Methylene blue	922	659
New methylene blue N	927	663
Indigotin (indigo; indigo carmine)	1,180	877
Prussian blue	1,288	968

VIOLET DYES

Methyl violet B	680	515
Crystal violet	681	516
Acid violet	698	530

BROWN DYES

Bismarck brown R	332	284
Benzamine brown 3GO	596	476
Benzo brown G	606	485

BLACK DYES

Durol black B	307	265
Columbia fast black FF (diamine fast black FF)	539	436
Direct deep black RW	582	463
Nigrosine base	864	698
Nigrosine, water-soluble	865	700

In the specifications for the two iron gallotannate inks, the blue dye that must be used in the standard is the particular soluble blue designated as C. I. No. 707. Schluttig and Neumann used the other soluble blue in the table, C. I. No. 705, which they called Bavarian blue DSF. The basic dye from which both soluble blues are derived is almost insoluble in water, but by treatment with strong sulphuric acid it is sulphonated to varying degrees, according to the conditions. The soluble blue of the specification is supposed to consist almost entirely of the trisulphonic acid or, rather, of its sodium salt. Bavarian blue is the sodium salt of the disulphonic acid, and it should therefore have a somewhat greater tendency than the trisulphonate to be precipitated by the tannin and the iron in the ink. Both tannin and iron salts are used as mordants in dyeing, because they form insoluble complexes with many dyes. Forty-two years ago, when Schluttig and Neumann selected Bavarian blue, it may have been the trisulphonate. Because some factory batches of the soluble blue turn out to be unsuitable for making ink, it is advisable when buying it to order it " for ink."

A few other blue dyes that are fairly satisfactory substitutes for soluble blue are naphthol blue-black S (C. I. 246; Sch. 217), benzo blue 2B (C. I. 406; Sch. 337), diamine sky blue FF (C. I. 518; Sch. 424), and benzo sky blue (C. I. 520; Sch. 426). Indigo carmine (C. I. 1,180; Sch. 877), which was the first dye used for the purpose, must not be overlooked.

In addition to Bavarian blue, Schluttig and Neumann named three other dyes, so that they could exactly match the shade of any sample submitted to them for test. Their red dye was azorubine, which they called nacarat S (C. I. 179; Sch. 163). Their "chestnut brown" can not be identified, for the only coloring matter listed under that name in the Colour Index is umber, which is not a dye, but an insoluble earth, used as a pigment. The third dye was acid green VBSPo or, it is now called, Guinea green B (C. I. 666; Sch. 502).

In making dyes it is sometimes necessary to salt them out of solution; that is, to precipitate them by dissolving common salt, usually, in the concentrated solution of the dye. When made in this way the dye unavoidably contains more or less salt. Many dyes also are intentionally mixed with salt or other uncolored substance to dilute them to the strength to which the dyer has been accustomed for decades. This is a recognized trade practice that is not to be regarded as adulteration. For making inks it is preferable to use the concentrated forms of the dyes, and these should be ordered from the manufacturer.

4. LITERATURE ON INKS

In this circular only a small percentage of the available formulas for inks have been given. Many more can be found in books that every public library, no matter how small, is likely to have, while there are other books that the larger libraries probably have. A few books of each class are in the list which is here given. In addition, three or four books on testing inks are mentioned.

1. Allen, A. H., Commercial organic analysis, 4th ed., vol. 5, pp. 669–690; vol. 9, pp. 456–468, P. Blakiston's Son & Co., Philadelphia, 1911 and 1917.
2. Brannt, W. T., and Wahl, W. H., Techno-chemical receipt book, H. C. Baird & Co., Philadelphia, 1905.
3. Bromley, H. A., Outlines of stationery testing, Chas. Griffin & Co., (Ltd.), London, 1913.
4. Carvalho, D., Forty centuries of ink, Banks Law Publishing Co., New York, 1904.
5. Henley's twentieth century book of recipes, formulas, and processes. Edited by G. D. Hiscox. Norman W. Henley Publishing Co., New York, 1928.
6. Lehner, S., Manufacture of inks. Translated, with additions by W. T. Brannt. H. C. Baird & Co., Philadelphia, 1892.
7. Mitchell, C. A., and Hepworth, T. C., Inks, their composition and manufacture, 3d ed., Chas. Griffin & Co. (Ltd.), London, 1924.
8. Osborn, A. S., Questioned documents, Lawyers' Cooperative Publishing Co., Rochester, N. Y., 1910.
9. Oyster, J. H., Spatula ink formulary, Spatula Publishing Co., Boston, 1912.
10. Schluttig O., and Neumann, G. S., Die Eisengallustinten (The Iron-Gall Inks), v. Zahn & Jaensch, Dresden, 1890.
11. Scientific American Cyclopedia of Formulas. Edited by A. A. Hopkins, Munn & Co. (Inc.), New York, 1921. Based upon an older book: Scientific American Cyclopedia of Receipts, Notes and Queries.
12. Spon, E., Workshop receipts, E. & F. N. Spon, London and New York, 1917.

13. Underwood, N., and Sullivan, T. V., Chemistry and technology of printing inks, D. Van Nostrand Co., New York, 1915.
14. Wiborg, F. B., Printing ink, Harper & Bros., New York, 1925. '

Two important articles that appeared in chemical journals are:

Munson, L. S., The testing of writing inks, J. Am. Chem. Soc., vol. 28, pp. 512–516; 1906.
Rupert, F. F., Examination of writing inks, Ind. Eng. Chem., vol, 15, pp. 489–493; 1923.

A few publications of the Bureau of Standards relate to inks. The Federal specifications, formerly issued as circulars of the bureau, are now parts of the Federal Standard Stock Catalogue. Many public, college, and university libraries throughout the country have the publications of the bureau, and possibly also the specifications. Those which are in print can be bought from the Superintendent of Documents, Government Printing Office, Washington, D. C., for the prices stated. Postage stamps will not be accepted in payment and money is at the sender's risk. Postal money orders, or coupons sold by the Superintendent of Documents in sheets of 20 for $1 are accepted. In ordering, the name of the bureau as well as the title and number of the publication should be given. The letter circular in the list is mimeographed, and is not handled by the Superintendent of Documents, but orders for all the other publications should be addressed to him.

J. B. Tuttle and W. H. Smith, Analysis of printing inks, B. S. Tech. Paper No. 39. 1915. (Out of print.)
Composition, properties, and testing of printing inks, B. S. Circular No. 53, 1915. (Out of print.)
Inks, typewriter ribbons, and carbon paper, B. S. Circular No. 95, 2d ed., 1925. (Out of print.)
P. H. Walker, B. S. Miscellaneous Paper No. 15, Some technical methods of testing miscellaneous supplies, 1916. A reprint, with notes and corrections, of Bur. Chemistry, Dept. Agric., Bulletin No. 109, 1912. 15 cents. The methods are not the same in all respects as those in the Federal specifications.
Dry etching of glass, B. S. Letter Circular No. 150. Free.
B. L. Wehmhoff and D. P. Clark, Standard mimeograph ink and paper, Govt. Ptg. Office Technical Bulletin No. 15, 1932. Free.

The following Federal specifications cost 5 cents apiece:

TT–I–521, Ink, copying and record.
TT–I–528, Ink; drawing, waterproof, black.
TT–I–542, Ink, marking, indelible (for) fabrics.
TT–I–549, Ink, red.
TT–I–556, Ink, stamp-pad.
TT–I–563, Ink, writing.

WASHINGTON. October 29, 1932.

O

CPSIA information can be obtained
at www.ICGtesting.com
Printed in the USA
BVHW04*1204060818
523683BV00013B/189/P